TRANSFORMING ANIMALS

TURNING INTO A JELLYFISH

by Tyler Gieseke

Cody Koala

An Imprint of Pop!
popbooksonline.com

abdobooks.com
Published by Pop!, a division of ABDO, PO Box 398166, Minneapolis, Minnesota 55439.
Cody Koala™ is a trademark and logo of Pop!.

Printed in the United States of America, North Mankato, Minnesota

102021
012022

THIS BOOK CONTAINS RECYCLED MATERIALS

Cover Photo: Shutterstock Images
Interior Photos: Shutterstock Images, 1–9, 18–21; Dennis Kunkel Microscopy/Science Source, 10; David Wrobel / BluePlanetArchive.com, 13–14; ALEXANDER SEMENOV/Science Source, 17

Editor: Elizabeth Andrews
Series Designers: Laura Graphenteen, Victoria Bates

Library of Congress Control Number: 2021942434

Publisher's Cataloging-in-Publication Data
Names: Gieseke, Tyler, author.
Title: Turning into a jellyfish / by Tyler Gieseke
Description: Minneapolis, Minnesota : Pop!, 2022 | Series: Transforming animals | Includes online resources and index.
Identifiers: ISBN 9781098241186 (lib. bdg.) | ISBN 9781098241889 (ebook)
Subjects: LCSH: Jellyfishes--Juvenile literature. | Marine animals--Juvenile literature. | Animal life cycles--Juvenile literature. | Marine invertebrates--Metamorphosis--Juvenile literature. | Animal Behavior--Juvenile literature.
Classification: DDC 593.5--dc23

Hello! My name is

Cody Koala

Pop open this book and you'll find QR codes like this one, loaded with information, so you can learn even more!

Scan this code* and others like it while you read, or visit the website below to make this book pop.

popbooksonline.com/turn-jellyfish

*Scanning QR codes requires a web-enabled smart device with a QR code reader app and a camera.

Table of Contents

Chapter 1

Transforming Animals

Jellyfish are animals that float through the world's oceans. They come in many sizes and colors. Jellyfish are famous for their tentacles. The tentacles sting!

Watch a video here!

Jellyfish are **transforming** animals. They grow through six steps. The steps are egg, planula, polyp, strobila, ephyra, and medusa.

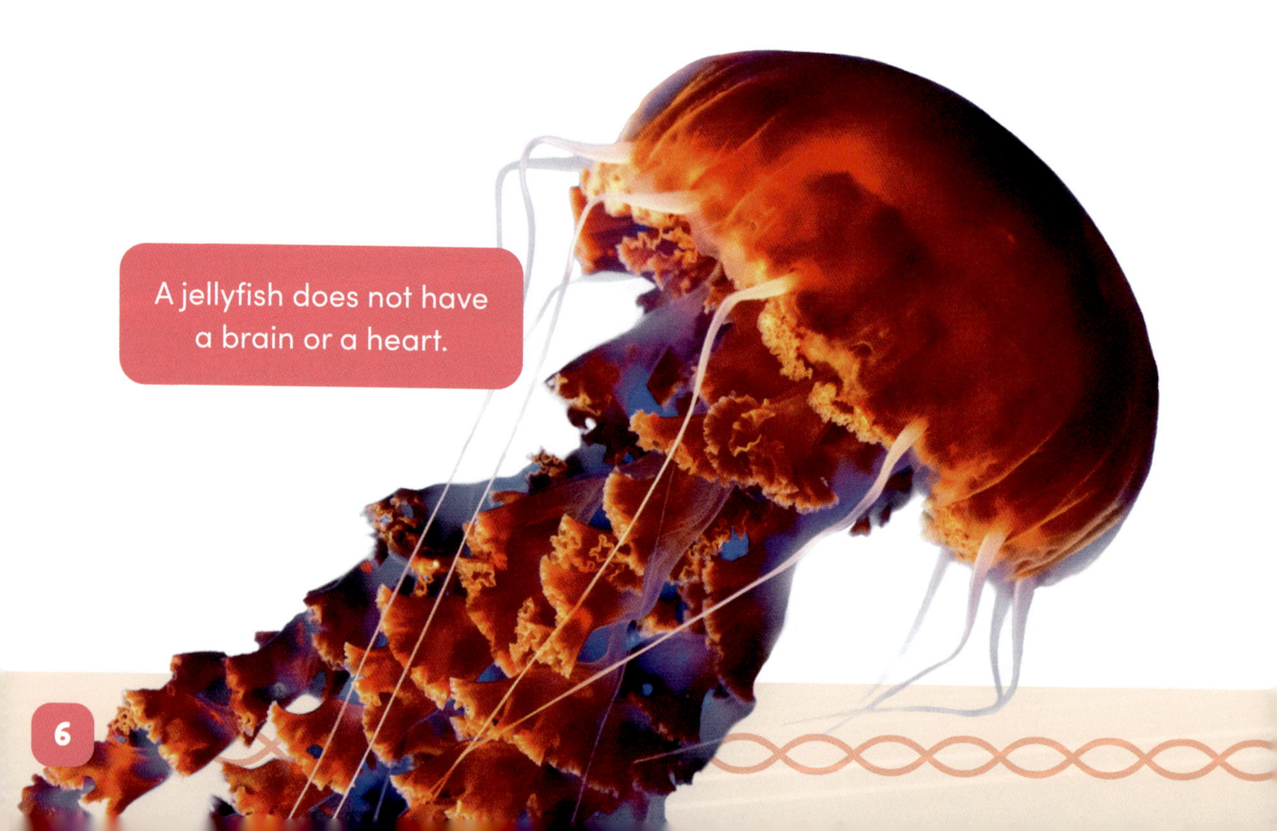

A jellyfish does not have a brain or a heart.

Life Cycle of a Jellyfish

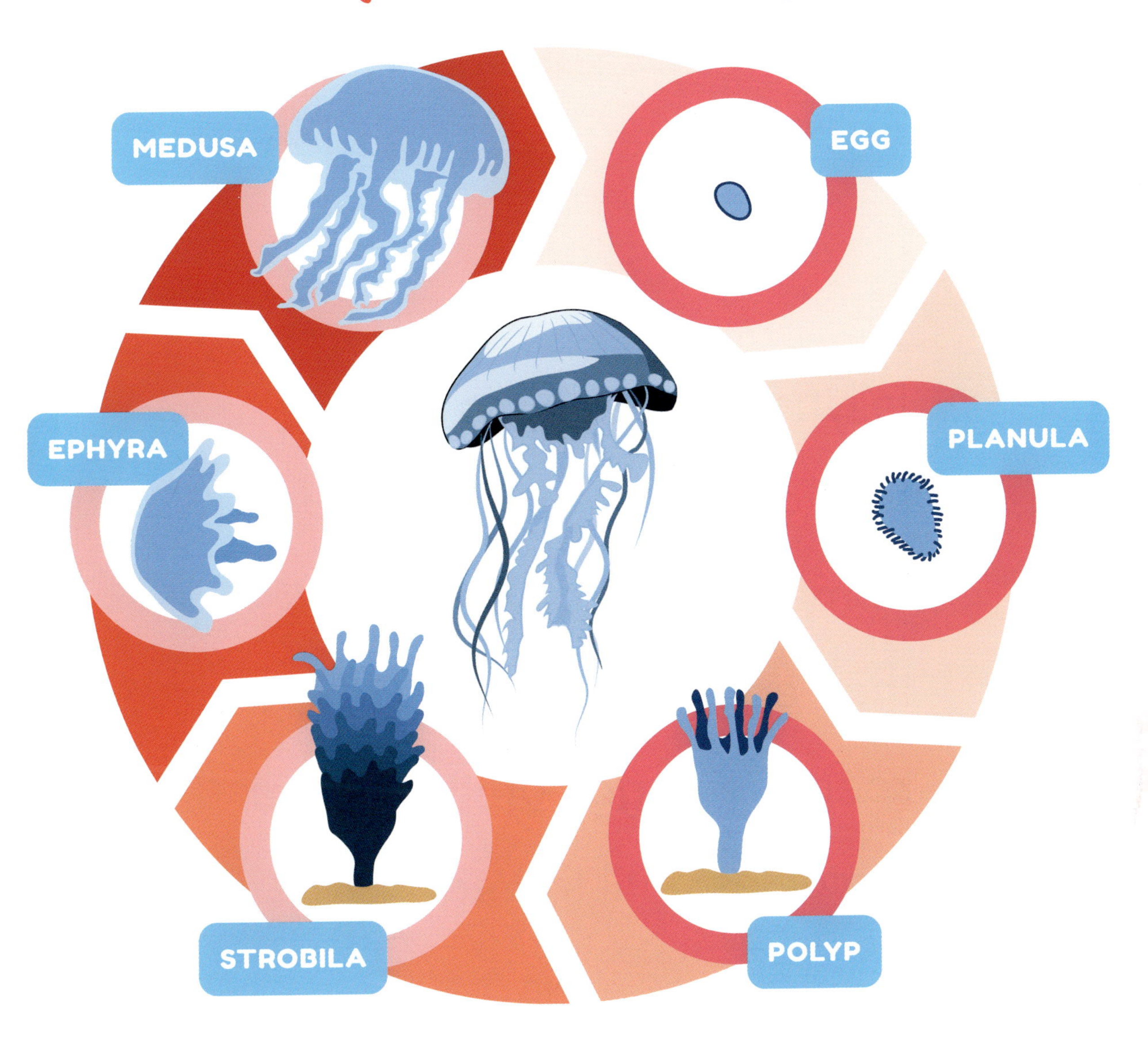

Chapter 2

Egg and Planula

Jellyfish create eggs when they are in large groups. Some female jellyfish let their eggs **drift** away. Others keep their eggs on their inner tentacles until the eggs **hatch**.

Learn more here!

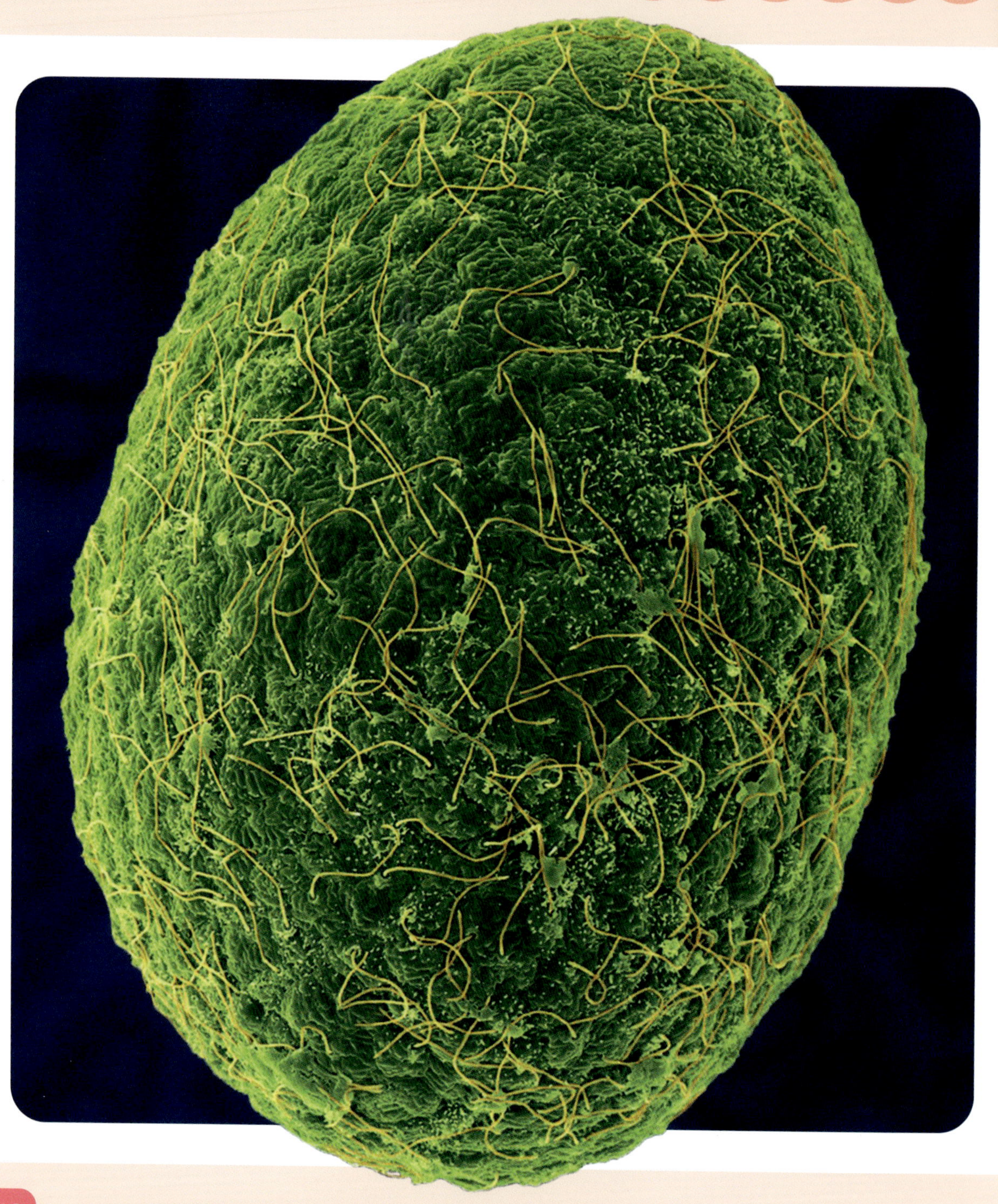

A newly hatched jellyfish is called a planula. It is small and looks flat.

The planula drifts for a few days. Then it attaches to something hard, like a rock. There, it grows into the next step of the life cycle.

Jellyfish were in the oceans before dinosaurs walked the earth.

Chapter 3

Polyp and Strobila

The planula is now a polyp. Polyps look like tubes with tentacles on the end.

A polyp pushes **plankton** into its mouth to eat. It can live in this stage for years.

Learn more here!

polyp
polyp

Some types of polyps turn right into adults. Other polyps split and become two adults.

Another type grows into a stack of 10 to 15 polyps. The stack is called a strobila. It looks like a tower.

Chapter 4

Ephyra and Medusa

One of the polyps breaks off a strobila. Now, it is called an ephyra. It is very small and **delicate** at first. The ephyra gets stronger as it grows into a medusa.

Complete an activity here!

bell
mouth
tentacles

A medusa is an adult jellyfish. A medusa has a **bell**, tentacles, and a mouth.

Many small jellyfish eat **plankton**. Large jellyfish can eat fish, shrimp, or even smaller jellyfish.

A medusa lives for just a few months. It finds a **mate**. The two jellyfish make new eggs. The life cycle starts again! Jellyfish are amazing **transforming** animals.

Making Connections

Text-to-Self

Which is your favorite step in the jellyfish life cycle? Why?

Text-to-Text

Have you read other books about sea animals? How are these sea animals similar to and different from jellyfish?

Text-to-World

Just one polyp can become a strobila and make many medusas. How does this help these jellyfish survive over time?

Glossary

bell – the round top of a jellyfish.

delicate – easily broken.

drift – to sit in and be moved around by water or wind.

hatch – to break out of an egg.

mate – a partner animal of the same kind. Together they make new eggs or babies.

plankton – very tiny water animals or plants.

transform – to change into a new shape.

Index

Online Resources

popbooksonline.com

Thanks for reading this Cody Koala book!

Scan this code* and others like it in this book, or visit the website below to make this book pop!

*Scanning QR codes requires a web-enabled smart device with a QR code reader app and a camera.